Famous caves of Europe:

5. Pennines, UK
6. Mendip Hills, UK
7. South Wales, UK
8. Massif Central, France
9. Dolomites, Italy
10. Dalmatia, Yugoslavia
11. Northern Spain

Famous caves of Asia and Pacific:

12. Oman
13. Kwangsi Province, China
14. Gunung Mulu caves, Sarawak, Malaysia
15. Java
16. Papua New Guinea

Famous caves of Australasia:

17. Nullabor Plain, South Australia
18. Hinchinbrook, Queensland
19. Waitomo 'Glow-worm' Caves, North Island, New Zealand

Famous caves of Antarctica:

None

3

FACTS ABOUT CAVES

The world's biggest limestone cave network so far discovered is Mammoth Cave National Park, Kentucky, USA. It extends for over 330 mi of explorable passageways.

The largest cave chamber in the world is the Sarawak Cavern in Mulu National Park, Sarawak, Malaysia. It is 2500 ft long and an average of 1000 ft wide and at least 200 ft high at every point.

The world's longest recorded hanging limestone column, or stalactite is in Malagar, Spain. It is 195 ft long. The tallest rising column, or stalagmite is at Lozere in France and is nearly 100 ft tall.

The world's deepest known caves are in France, extending down nearly 5000 ft.

Although the largest caves are usually made from limestone, caves can also form in other rock types. Many small caves are etched by the action of the waves and the wind. Caves and tunnels are also common in and under glaciers (*for more information see the book* Glacier *in the Landshapes set*) and they sometimes form in the lava that flows from a volcano. Lava Beds National Monument in California, USA, has a 80 square mile area of underground lava tubes. (*For more information on lava see the book* Volcano *in the Landshapes set.*)

 Grolier Educational Corporation
SHERMAN TURNPIKE, DANBURY, CONNECTICUT 06816

LAND 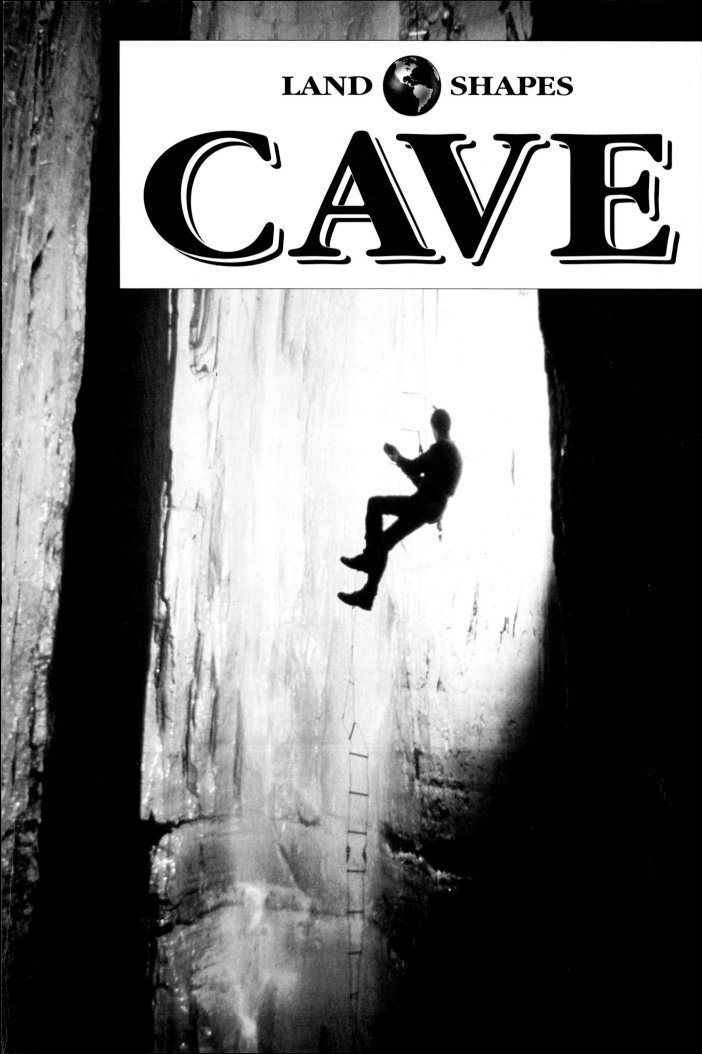 SHAPES

CAVE

Author
Brian Knapp, BSc, PhD
Art Director
Duncan McCrae, BSc
Editor
Rita Owen
Illustrators
David Hardy and David Woodroffe
Print consultants
Landmark Production Consultants Ltd
Printed and bound in Hong Kong
Designed and produced by
EARTHSCAPE EDITIONS

First published in the USA in 1993 by
GROLIER EDUCATIONAL CORPORATION,
Sherman Turnpike, Danbury, CT 06816

Copyright © 1992
Atlantic Europe Publishing Company Limited

Library of Congress #92–072045

Cataloging information may be obtained
directly from Grolier Educational Corporation

Title ISBN   0–7172–7184–6

Set ISBN   0–7172–7176–5

Acknowledgements. The publishers would like
to thank Redlands County Primary School.

Picture credits. All photographs from the
Earthscape Editions photographic library except
the following (t=top, b=bottom, l=left, r=right):
David Higgs 14; Tony Waltham Geophotos *Cover*, 5,
8/9, 19t, 19b, 31, 32/33, 34/35, 35t.

Cover picture: Gaping Gill Cave, UK.
Inside back cover: Lehman Caves, Nevada, USA.

In this book you will find some
words that have been shown in **bold**
type. There is a full explanation of
each of these words on page 36.

On some pages you will
find experiments that you
might like to try for
yourself. They have been
put in a blue box like this.

In this book mi means miles and
ft means feet.

These people appear on a number
of pages to help you to know the
size of some landshapes.

CONTENTS

Introduction

Caves are part of a world that many of us never see. Yet below the surface there are chambers bigger than any concert hall, tunnels wider than a multi-lane highway and rivers that can be raging torrents. Within these caves can be found spectacular columns and shapes like icicles decorating many of the chambers and creating some of the most beautiful places on Earth. Each of the features has been produced because a commonplace rock they form in – called **limestone** – can be slowly dissolved by the world's most common liquid – water.

Caves are dark and mysterious. They still offer a sense of wonder and excitement to everyone who enters them, whether as a tourist walking through a well-lit cave system or an amateur cave explorer (called a speliologist or spelunker) crawling through the tightest of passageways such as the one shown here.

Caves have been formed over millions of years as the cracks in the limestone have slowly been widened by water. They are fascinating, not just because there are large caverns, but because caves are linked together by passageways into huge systems that may stretch for hundreds of miles.

In this book you can find out about the hidden world of the cave, about the many beautiful and delicate features that are found inside them, and about how all these features have been formed. Enjoy finding out about caves by turning to a page of your choice.

Care in the caves

Caves are exciting and mysterious places to be in, but they are completely dark and commonly have deep pits and unstable roofs. The caves that are open to the public are usually safe and contain many of the best formations. Never enter even the smallest 'unofficial' cave without an experienced and 'qualified' guide who knows the cave well. People have been killed because they have ignored this advice.

Care for the caves

Caves form slowly over many thousands of years. There are delicate formations which can easily be damaged. Do not touch the formations because even the sweat from your skin may cause damage. People in the past have broken pieces of stalactite from the caves and in doing so they have caused sadness to many millions who followed. So never collect samples from a cave. All of the specimens photographed in this book have been taken from scientific collections: none have been collected just to keep at home.

Chapter 1:
How caves form

The special nature of limestone

All of the world's largest caves form in a rock called limestone.
Limestone is a very tough rock but it has one important weakness;
it will **dissolve** very slowly in water, it can be carried in **solution** by
running water, and it is easily deposited again layer on layer to form
spectacular shapes. This is the key to understanding why caves and
all other limestone features occur.

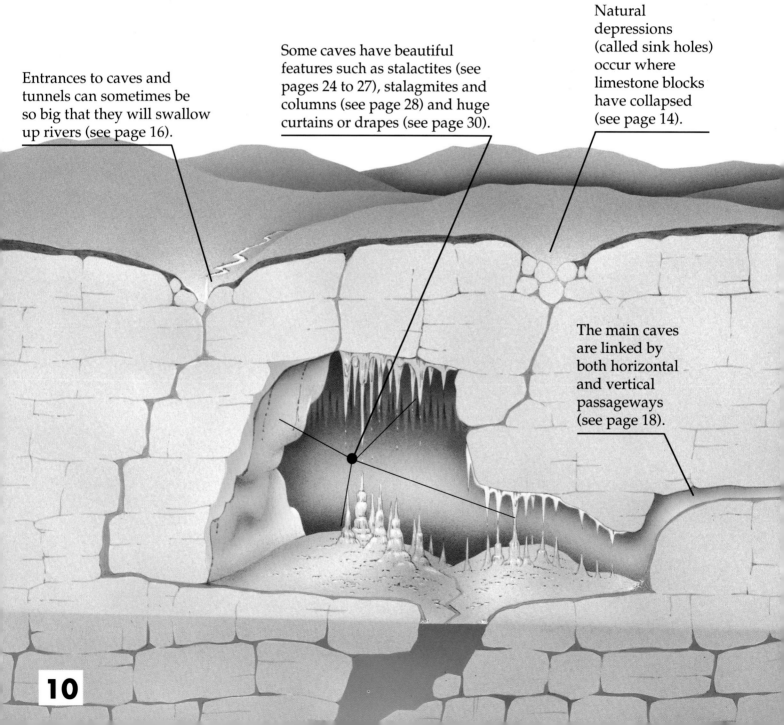

Natural
depressions
(called sink holes)
occur where
limestone blocks
have collapsed
(see page 14).

Some caves have beautiful
features such as stalactites (see
pages 24 to 27), stalagmites and
columns (see page 28) and huge
curtains or drapes (see page 30).

Entrances to caves and
tunnels can sometimes be
so big that they will swallow
up rivers (see page 16).

The main caves
are linked by
both horizontal
and vertical
passageways
(see page 18).

This is a piece of limestone. It comes in many colors from pink through white to grey. Massive pieces of limestone make up the natural blocks that are slowly dissolved to make caves.

The way the water gets underground depends on patterns of hairline cracks in the rock (see page 12).

Caves form in limestone that is made of very large blocks (see page 20).

Gorges show the natural shape of the limestone blocks.

Rubble on the floor of a cave shows where roof falls have occurred.

Water often emerges at the foot of a limestone cliff (see page 22).

Dissolving pavements

Clints – blocks – and **gryke**s – cracks – are two strange-sounding words that people use in limestone country.

In some parts of the world large areas are covered with clints and grykes, making the land look as though it is covered in paving slabs. But these are no ordinary **pavements**; their curious shapes hold the clue to cave formation.

Dissolving limestone

This is an area of bare limestone pavement that was once covered with soil. During the **Ice Age** it was stripped away by advancing ice sheets and there has been no time for new soil to develop since.

Soil contains roots of plants and many small animals. Just like people, each of them gives off a gas called **carbon dioxide** as they 'breathe'. Carbon dioxide gas and water combine to make an acid (**carbonic acid**) that can attack the rock, slowly but surely eating it away. This picture shows how the cracks, or grykes, are made larger as limestone rock is attacked by carbonic acid.

This picture shows a part of the of limestone pavement that has been etched by water over thousands of years. The water has seeped down the cracks and dissolved some of the limestone, so the cracks are now quite wide.

The same process is also widening cracks deep underground.

How cracks widen

Grykes, cracks that are found on the top of a limestone cliff, have already been widened by carbonic acid. However, the cracks you see on a fresh cliff **face** are still little more than hairline thickness and have yet to be widened.

Much of the water seeps through the rock along hairline cracks, gradually dissolving away the rock. Eventually some of the cracks become wide enough to let water flow quickly. As soon as this happens tunnels and caves can form.

The way water widens cracks in limestone can sometimes be seen in the most unusual way. Here a climber is using the cracks that have been widened by solution as places to hammer home small metal wedges called pitons. By suspending himself from the pitons he can traverse the cave roof upside down!

Here a horizontal crack has been enlarged enough for water to be able to pour through it.

The cracks that criss-cross a limestone rock can be clearly seen on this limestone cliff.

As cracks widen, the limestone blocks are loosened and eventually many fall on to each other leaving a jumble of rock. This is the reason large caves are rare landshapes.
  The caver in the picture below is squeezing his way between some collapsed blocks of limestone.

Disappearing rivers

The size and shape of a cave system depends, in part, on how water gets into the limestone. Only a few caves have wide, gaping entrances, most have entrances that are little more than modest holes in the ground. In some places the ground even appears to swallow rivers through no obvious entrance.

Like a great open mouth waiting to swallow the water, this cave entrance on the Popo Agie River in Wyoming, USA, takes the entire flow of the river. The cave entrance, and the tunnel leading into the rock, have been dissolved away over thousands of years. If the water finds a way through the limestone farther upstream, a new entrance will be formed and this cave will eventually become dry.

Gaping hole

In some places rivers run over the land surface and then disappear down deep well-like shafts. These 'swallowing holes' are especially common where the limestone is fractured with horizontal and vertical cracks. When water finds a crack where the limestone is easy to dissolve then it is made bigger and bigger until a 'well' forms. Some of these entrance wells, such as Gaping Gill, shown here and on the cover of the book, are hundreds of feet deep.

Passageways

Limestone rocks often contain a **labyrinth** of tunnels. In general the pattern of tunnels criss-cross at right angles, following the original directions of the cracks. Cavers following these tunnels usually make slow progress because the tunnels are often small and twist and turn every few yards. Occasionally, tunnels are wide enough for people to stand up in, and can be straight for many yards. These unusual, but spectacular, tunnels are the ones that have often been made into show caves and opened to the public. Here are some shapes of 'walk-through' tunnels and shafts and the reasons for their shapes.

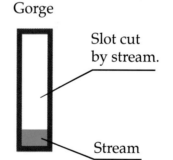

Gorge

Slot cut by stream.

Stream

Many of the upper passageways in limestone have been cut by running streams. The gorges they have formed are deep and narrow. The picture above shows one such gorge.

Pipes and gorges

Tunnels and shafts either have rough or smooth sides. If they are smooth walled they were formed when water completely filled the tunnel and they were, in effect, pipes. Such pipes are also round in cross-section, just like a drainpipe.

If tunnels are rough walled then water flowed only in the bottom, cutting mini-gorges into the rocks as you can see in the picture on the left.

This upper 'pipe' shape was cut when water filled the tunnel.

This lower 'gorge' shape was cut when a stream flowed in the bottom of the tunnel.

Keyhole tunnel.

Cut by stream when it completely filled the pipe.

Cut by stream when the pipe was no longer completely filled with water.

A large stream once filled this shaft completely. Today water still falls from a passageway high in the wall.

Keyhole tunnels

Some tunnels (such as the one shown in the picture above) are 'keyhole' shaped. They are formed in two stages.

As water rushes through a tunnel that is completely filled with water, it scours the tunnel into a near perfect tube. If, however, the tube is partly drained as water finds a new, lower passage through the rock, only the section of the tunnel with flowing water will be scoured, eventually cutting a gorge in the bottom of the tunnel. Soon the tunnel becomes keyhole-shaped.

Shafts

Shafts are vertical tunnels. The water that tumbles vertically through a shaft has enormous energy and so it can create large, wide routes. Many such shafts have waterfalls spilling down their sides.

Caverns

A cavern, is a large underground chamber. Some chambers can be taller than cathedrals, wider than main highways and cover more ground than a stadium. But most are quite modest in size, perhaps 30 ft high, 150 ft long and 30 ft wide.

Chambers are connected by passageways, and often several passageways connect into each chamber. The most important thing is that chambers are places where water can drip from the ceilings, ooze down the walls and spread over the floors. This is the key to the growth of the beautiful formations for which caves are famous.

Spectacular cave formations, such as these at Lehman Caves, Nevada, USA, require very special conditions to form in (see page 24).

Cave decorations are a rarity. In the great majority of cave systems the right conditions for producing these spectacular formations just do not exist.

Formations take hundreds of thousands of years to form. This cave is only young and it is open to the air, so it has no large formations.

Where water surfaces

The water that flows into a region of limestone rock is certain to emerge at the surface again. In some places it will start a 'new' river at the foot of a cliff like the ones shown on these pages.

Here, all the water that first seeped through tiny cracks, then gathered together to make underground rivers, appears as if by magic at the foot of a limestone cliff.

The rock wall of a cliff may hide limestone that is riddled with tunnels and caves. Just a few yards away from the emerging river, an underground tunnel may be filled with rushing water like that shown in the picture above.

From time to time, as the cliff wears back, parts of the underground system are exposed, sometimes as tunnels, at other times as caves.

This exit from a cave system has been widened both by solution and the scouring power of the flowing water. This is the reason that rivers leaving limestone rocks often make caves.

The water flowing out at the foot of this cliff at Malham in England has been gathered from an area of hundreds of square miles of limestone rock. The cracks, tunnels, shafts and caves within the limestone store water after each storm and release it gradually. As a result, this stream will be able to keep flowing even during a drought.

Cliff made of limestone.

Water emerges at the foot of the cliff.

23

Chapter 2
Cave features

Drip, drip, drip

The lime that is dissolved in water only stays in solution when it is in tiny cracks. As soon as water seeps out into a cave it drips from the roof but the lime stays behind. This is why.

Why drips form lime

Carbon dioxide in the water that seeps through the limestone rock produces a weak carbonic acid that slowly dissolves the rock.

While the limestone-rich waters are trapped inside water-filled cracks, the carbon dioxide stays locked in the water. But when it reaches a cave the gas bubbles away and the lime is left behind as a deposit.

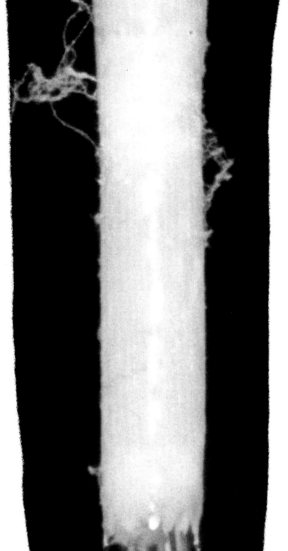

This is a drop of water on the end of a rod of lime scale. A drip falls off this rod about every minute; this rod grows about 1 inch every 50 years!

As each drop falls away a tiny ring of limestone is left behind. Slowly but surely the rings create vertical tubes of limestone called soda straws. If a draft catches the water drips as they hang, the scale will grow away from the vertical and it can cause some weird spaghetti-like shapes to form.

These dark lines are the cracks between the blocks of limestone. Notice how deposits are building up around them as water seeps out.

Find the hidden limestone

The water in your tap may have many hidden minerals dissolved in it. To make them appear simply place a transparent dish in the sunlight and partly fill it with tap water.

Leave it for a few days and when the water has gone and the dish is dry, there will be a white film on the inside of the dish. Most of this film will be limestone.

The inside of a kettle can show quite thick deposits of limestone. They form because hot water can hold less dissolved limestone that cold water.

With the kettle unplugged and empty, carefully try to scrape the limestone off the inside of the kettle with a spoon. You will soon see how tough the limestone is.

These pictures show in close-up some of the strange soda straw shapes. Some people have likened the shapes to sea-horses.

Stalactites

These are rods of limestone that hang down from the roof of a cave. They form some of the most spectacular shapes inside a cave. The largest can be many yards long and can weigh many tons, yet they rarely fall.

Stalactites start life as a tiny straw, similar to those shown on the previous page. If the tube becomes blocked by limestone deposits (scale), the water flows down the outside of the tube, causing the base to thicken.

Stalactites can only form where cave roofs are more or less level. If the cave roof slopes steeply, the drops of water will simply run down the walls.

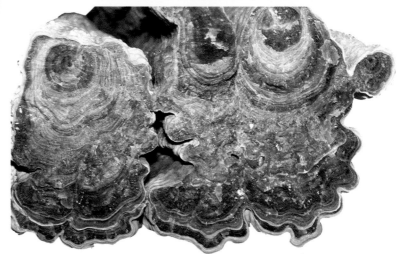

This cross-section of a stalactite clearly shows a layered growth pattern, much like the rings you see on a cross-section of a tree. However, these rings take much longer to form than the growth rings on a tree. Each dark and light band indicates the changes in the amount of iron that was present in the water. Each ring can take tens or even hundreds of years to form.

A group of stalactites have formed from a single cave roof leak. Now they hang from the ceiling like giant spears.

Stalagmites and columns

Stalagmites are rods of lime that grow upwards from the floor of a cave. They are nearly always thicker than stalactites. The reason for this is that water flows *to* the tip of a stalactite, making it longer, whereas water flows *from* the tip of the stalagmite, making it broader.

The picture on the right shows a close-up of where a stalactite (above) meets a stalagmite (below) to make a column.

How stalagmites grow

The stalagmite in this picture looks like a many-layered cake covered in icing. This is because the stalagmite has grown in stages.

Remember that stalagmite rods grow upwards from the floor of a cave where water drips from the roof. As the rod grows, the water is always flowing back down the rod, at the same time leaving behind a deposit of lime. This is why stalagmites tend to be stump-like.

The stumps tend to broaden and form flat tops. Each tier of the stalagmite represents a stage of broadening. Each set of 'icicle' stalactites indicates where the water dripped off the broadening stump.

Stage of growth

Broadening stump

Mini-stalactites

How columns form

As water drips from the ceilings, so the lime scale often blocks the cracks through which the water seeps, and this stops the stalactite and stalagmite from growing. As water flows first from one crack, then from another, stalactites grow all over the roof, and stalagmites all over the floor.

Only rarely, in this merry-go-round of leaks does the water continue to flow from the same spot for thousands of years. But when it does it makes columns, often the most beautiful of all the cave forms.

The picture on the left shows a slender column in Nevada's *Lehman Caves*.

Flowstone

It is only possible for stalactites and stalagmites to form when the roof of the cave is fairly level and when drops of water fall from the roof. But in many instances water simply seeps out along a crack on a sloping roof and thus runs down the wall where it then builds a limestone scale which is called **flowstone**. As a result the walls of the cave become masked in a sheet of smooth limestone.

The picture above shows how flowstone can build up in thick layers until it completely masks the original shape of the cave wall.

Quite often flowstone will form a number of thin vanes sticking out from the wall of the cave. The flowstone in the picture to the right has formed shapes rather like the drapes that hang beside a window.

Curtains or rods?
A bewildering variety of shapes are formed in caves. So what makes the limestone become a curtain or drape rather than a stalactite, or visa versa?

Try this experiment. Get a sheet of some porous solid (such as a sheet of foam) and put it as flat as you can. Now add water to the top and watch how the water comes out of the bottom.

Very slightly tilt the foam and see what happens to the way the water comes out.

River flows over here.

This caver is using a rope to climb up a 'waterfall' of limestone. This curtain flowstone is still forming and from time to time a stream still makes a waterfall over the lip of the flowstone. The flowstone has been built up directly from the flowing waters of the stream.

Chapter 3:
Caves of the world

Carlsbad Cavern, USA

Carlsbad Caverns National Park is part of a huge limestone plateau in New Mexico, USA. The chambers in the caves system are called caverns because they are so big.

They were formed by dissolving an ancient coral reef about 600 ft below the plateau surface. Visitors can walk for over 3 miles down through a series of caverns that may have had their origins 60 million years ago, before they return to the surface by elevator.

Much of the water that seeped through Carlsbad's limestones was heated, and made more acid, by volcanic rocks beneath the limestone. These hot, acid waters are one reason that Carlsbad has such enormous caverns.

This picture shows the entrance to the caverns. Just inside the entrance an estimated one third of a million Mexican freetail bats are known to roost!

Even main passageways are of gigantic proportions as you can see in the picture above. The pathway wall is about three feet high.

How the caverns formed

The caverns began dissolving away when the reef rocks were still filled with water. It was only about a million years ago that the water level fell and the caves began to dry out. Since then flowstone, stalactites and stalagmites have begun to form in most of the chambers and passageways.

This is the entrance to the Big Room in Carlsbad Caverns. It is one of the largest known underground chambers in the world, extending for 60 acres and over 60 ft high.

Mulu caves, Malaysia

The world's largest caves lie hidden beneath the tropical rainforests of Sarawak, a part of Malaysia. The caves were completely unknown until 1978. Now they form part of the Gunung Mulu National Park. In the center of the park lies a chain of limestone mountains only 20 mi long, yet below the surface lies at least 60 mi of enormous caverns and passages carved in almost pure limestone rock.

Huge caves and vast tunnels
Mulu has one of the highest amounts of rainfall anywhere in the world. The rainfall gathers into large rivers which have carved spectacular passageways.

The most impressive passage in Mulu is called the Deer Cave (Gua Payau). It is half a mile long and nowhere is it less than 200 ft high and wide. It is home to over one million bats.

Good Luck Cave (Lubang Nasib Bagus) contains the world's largest cavern, the Sarawak Chamber. It is 2200 ft long, 300 ft high and 1200 ft across and many times larger than the Big Room in Carlsbad. It would hold up to 7500 coaches!

This is a typical giant passageway in Mulu.

Many of the entrances to the cave systems are overhung with rainforest vegetation. The picture on the right shows a river flowing into the throat of a cave.

New words

cave
a large underground network of passages and chambers formed by natural processes and which are big enough to be explored. Most caves are dissolved out of solid limestone by the slow action of water. Peolple sometimes use the word 'cave' as a shorthand for 'cave system'

clints and grykes
A clint is a block of limestone and a gryke is the widened crack between blocks

carbon dioxide
a gas which is found in the atmosphere. Carbon dioxide is also produced by animals living inside the soil. Rainwater absorbs carbon dioxide, forming a weak acid which can then dissolve limestone rock

carbonic acid
a chemical which is able to dissolve limestone and widen cracks between limestone blocks into passageways and caverns. Carbonic acid is formed as carbon dioxide gas is absorbed by rainwater

dissolve
a material which can be absorbed by a liquid without showing any visible signs of its presence. Rainwater contains many invisible substances that have been dissolved in its passage through air, soil and rock

face
the clear-cut edge of a rock

flowstone
the general name for all the deposits other than stalactites and stalagmites that have been formed as limestone comes out of solution. Most flowstone makes a kind of drapery on cave walls and ceilings

Ice Age
a time, beginning about a million years ago, when the Earth became colder and ice sheets expanded to cover nearly a third of the land. During this time ice sheets swept soil from many areas and left limestone rocks bare. Since the ice retreated a few thousand years ago there has not been time for new soil to form

labyrinth
a maze of twisting and turning underground passageways

limestone
a special rock made from a substance called calcium carbonate. Chalk is also calcium carbonate

pavement
a name given to large areas of exposed flat limestone rock where the widened cracks and blocks can easily be seen. The largest area of limestone pavement in the world is in western Ireland, although this is not a region of large caves

solution
the process whereby solid materials are gathered and dissolved in a liquid. Calcium carbonate is brought into solution when it reacts with weak carbonic acid

stalactite
a long tapering rod of limestone that hangs down from the roof of a cave at a place where water seeps out from cracks in the rock

stalagmite
a broad, tapering rod of limestone that grows up from the floor of a cave at a place where water drips from the roof

Index